In the Garden

by Pearl Markovics

Consultant:
Beth Gambro
Reading Specialist
Yorkville, Illinois

Contents

In the Garden. 2
Key Words 16
Index. 16
About the Author 16

New York, New York

What do you see?

I see carrots in the garden.

What do you see?

I see corn in the garden.

7

What do you see?

I see worms in the garden.

I see tomatoes in the garden.

What do you see?

I see potatoes in the garden.

What do you see in the garden?

Key Words

carrots 　　**corn**

potatoes　　**tomatoes**　　**worms**

Index

carrots 4–5　　potatoes 12–13　　worms 8–9
corn 6–7　　tomatoes 10–11

About the Author

Pearl Markovics grew up on a small farm in New York. She enjoys the sweet taste of garden fresh tomatoes.

Teaching Tips

Before Reading

- ✔ Guide readers on a "picture walk" through the text by asking them to name the things shown.
- ✔ Discuss book structure by showing children where text will appear consistently on pages.
- ✔ Highlight the supportive pattern of the book. Note the consistent number of sentences found on each page.

During Reading

- ✔ Encourage readers to "read with your finger" and point to each word as it is read. Stop periodically to ask children to point to a specific word in the text.
- ✔ Reading strategies: When encountering unknown words, prompt readers with encouraging cues such as:
  - **Does that word look like a word you already know?**
  - **Check the picture.**

After Reading

- ✔ Write the key words on index cards.
  - **Have readers match them to pictures in the book.**
  - **Have children sort words by category (words that end with the same four letters, for example).**
- ✔ Ask readers to identify their favorite page in the book. Have them read that page aloud.
- ✔ Ask children to write their own sentences. Encourage them to use the same pattern found in the book as a model for their writing.

Credits: Cover, © Oleg Romanko/Shutterstock and © Fedor Selivanov/Shutterstock; 2–3, © Phil Darby/Shutterstock; 4 5, © samael332/iStock; 6–7, © Shinyfamily/iStock; 8–9, © Mik122/iStock; 10–11, © ZolaKostina/iStock; 12–13, © nednapa/Shutterstock; 14–15, © FatCamera/iStock; 16T (L to R), © samael332/iStock and © Shinyfamily/iStock; 16B (L to R), © nednapa/Shutterstock, © ZolaKostina/iStock, and © Mik122/iStock.

Publisher: Kenn Goin **Senior Editor:** Joyce Tavolacci **Creative Director:** Spencer Brinker **Photo Researcher:** Thomas Persano

Library of Congress Cataloging-in-Publication Data in process at the time of publication (2019)
Library of Congress Control Number: 2018047592
ISBN-13: 978-1-64280-206-1 / ISBN: 978-1-64280-379-2 (paperback)

Copyright © 2019 Bearport Publishing Company, Inc. All rights reserved. No part of this publication may be reproduced in whole or in part, stored in any retrieval system, or transmitted in any form or by any means, electronic, mechanical, photocopying, recording, or otherwise, without written permission from the publisher. For more information, write to Bearport Publishing Company, Inc., 45 West 21st Street, Suite 3B, New York, New York 10010. Printed in the United States of America.

10 9 8 7 6 5 4 3 2 1